# BCBA & BCaBA Practice Exam

**Tony Mash**

HIGHLAND BOOKS


Tony Mash
BCBA & BCaBA Practice Exam

ISBN-13: 978-1500356613
ISBN-10: 1500356611

Printed in the United States of America

# BCBA & BCaBA Practice Exam

Number of Questions: 160
Time Limit: 240 minutes (4 hours)

1. Which of the following is an example of an operant extinction?
   a) You get a mosquito bite on your arm and it is itchy. You apply anti-itch cream on your arm, so your arm would not be itchy.
   b) When you are driving the car in the evening, the sun is glaring. You lower the visor so you can block the sunlight.
   c) Your daughter is always on the phone with her friends and the telephone bill is very expensive. You call the telephone company and cut her phone service. Now when she calls her friends using her phone, it is not working.
   d) You have a seasonal allergy and often sneeze. In the morning, you take medicine and it stops your allergic symptoms.

2. Your teacher yells at you when you talk to your classmates during class so you stay quiet when the teacher is in the classroom. Your teacher is a _____
   a) stimulus
   b) reinforcer
   c) punisher
   d) prompt

3. Signing "hello" when someone says "hello" is _____.
   a) textual
   b) echoic
   c) intraverbal
   d) taking dictation

4. When you are walking on the street, you see a house on fire. You yell "fire" and people in the house evacuate. Seeing a house on fire is a(n) _____.
   a) EO
   b) $S^D$
   c) reinforcer
   d) punisher

5. Which of the following is an example of a within-stimulus prompt?
   a) Whispering the answer
   b) Highlighting the answer
   c) Showing a gesture of the answer
   d) Pointing to the answer

6. Which of the following is true about Direct Instruction?
a) The teacher gives students questions and answers simultaneously.
b) Direct Instruction is an intense instructional method that requires the teacher to target only one student to teach him/her skills.
c) It is fast-paced.
d) This procedure was developed by Ogden Lindsley.

7. When Henry hit his sister, his father brought a pillow and told Henry to repeatedly hit the pillow until he was completely exhausted. This punishment procedure is known as a ______.
a) distributed practice
b) tolerance training
c) positive practice
d) negative practice

8. Mary is a BCBA working at an elementary school. One day, a special-ed teacher tells her that one of her students hits his own cheek. When Mary asks the teacher what the antecedents are, the teacher tells her that it just happens out of the blue. Mary then observes the student and investigates what is triggering his cheek hitting. Which attitudes of science does it contribute to?
a) Experimentation
b) Empiricism
c) Determinism
d) Philosophic doubt

9. When you want to confirm that your client is generalizing the skill that you are teaching, you should apply a(n) _____.
a) indiscriminable contingency
b) CRF
c) stimulus fading
d) most-to-least prompts

10. Which of the following best describes a behavior event?
a) Daniel was tired, so he rested on the couch for 10 minutes.
b) Sandy started biting her arm as soon as she saw a dog on the street.
c) Bill started crying because he was thinking about his grandfather who passed away a month ago.
d) Susan does not like darkness, so she always turns on all lights in her house.

11. Mike is a 4-year-old boy with Autism. He sometimes puts marbles in his mouth. When he does it, his mother yells at him and tells him to stop. However, Mike often giggles and enjoys seeing her reaction. Mike has never swallowed the marbles but his mother is concerned that he may swallow them someday. What would be the most significant and quickest intervention for this behavior?
a) The behavior is maintained by attention given by his mother. Since he is not intending to swallow them, she should place his behavior on extinction by ignoring it.
b) His behavior could be life threatening and Mike needs to realize how dangerous it is so his mother should continue telling him to stop.
c) As soon as Mike puts the marbles in his mouth, his mother should remove them from his mouth and make him engage in a non-preferred activity as a consequence.
d) His mother should hide all marbles and block access to them.

12. Catherine is a gym trainer at a gym. When she asks Richard to do 2 sets of 20 pushups, it takes 10 seconds for him to start. Richard takes 60 seconds to complete 20 pushups. He then takes a 3-minute break and starts the second set of 20 pushups. It takes 5 minutes and 10 seconds to complete 2 sets of 20 pushups after Catherine instructs Richard. What is the inter-response time in this scenario?

a) 10 seconds
b) 60 seconds
c) 3 minutes
d) 5 minutes and 10 seconds

13. Which of the following procedures allows a student to learn the material at their own pace and is known as the Keller Plan?

a) DTT
b) PSI
c) DISTAR
d) TEACCH

14. Eric has a drinking problem, so you, a BCBA, are asked to make a behavior plan to reduce his drinking. After working with him for several months, he successfully drinks less often and you make a behavior contract stating that Eric can only drink on Tuesday night and Friday night. One Tuesday night, he forgets that it is the day he can drink. What should you do as a BCBA?

a) You should remind him that he can drink on Tuesday.
b) You should avoid telling him that he can drink that day as it is his goal not to drink.
c) You should give him another type of reinforcer for not drinking.
d) You should wait and see if he would drink and if not, give it to him the following day.

15. Steve is a BCBA working with Edward, a 12-year-old boy with Autism. Edward engages in the self-injurious behavior of slapping his face on average of once every 15 minutes. After conducting a Functional Analysis, Steve concluded that the behavior was maintained by gaining attention from others. Which of the following would NOT be an effective intervention?

a) Teaching Edward to tap another person's shoulder to get his/her attention as a replacement behavior
b) Placing his behavior on extinction by not talking to him or making eye contact
c) Applying non-contingent reinforcement by providing him attention every 20 minutes
d) Using differential reinforcement of incompatible behavior by giving him praise when he is playing with a videogame and keeping both hands occupied.

16. Which of the following is an example of symmetry?

a) Upon seeing a picture of a quarter, Susan is taught that it is called "quarter". Susan then is taught that a quarter is 25 cents. When Susan sees a picture of a quarter, she says "25 cents."
b) When given an apple, Joshua matches it to another apple.
c) When shown pictures of a car, a truck, a motorcycle and an airplane, Sean says, "they are all vehicles."
d) A Spanish teacher shows his students a picture of a house and has students say "casa". When he says, "casa", his students point to the picture of a house.

17. Which branch of behavior analysis focuses on the philosophy of behavioral science?
   a) Experimental Analysis of Behavior
   b) Empiricism
   c) Applied Behavior Analysis
   d) Behaviorism

18. Which statement is NOT true about a graph?
   a) A bar graph always displays the variability of behaviors.
   b) A line graph can include more than one behavior and one condition.
   c) There are no data paths in scatterplots.
   d) Both a bar graph and a line graph are based on a Cartesian plane.

19. Which of the following is an example of a parametric analysis?
   a) When Alex was shopping he found two clocks he liked. One was $15 and the other one was $30. He had $20 so he chose to buy the $15 clock.
   b) When David weighed his bag it was 20lbs. He then knew it was 9kgs.
   c) Tina's dad told her that he would give her $5 for washing his car. Tina said no so her dad offered her $10. Tina then washed his car. Tina's dad realized that she would wash his car for $10 but not for $5.
   d) When a mother said, "we have bananas for snacks today so come here", Bill ran quickly to get one but his twin brother Scott did not come. The mother knew Bill liked a banana but Scott did not.

20. Gary is a 47-year-old man with Mild Intellectual Disability who lives in a group home. He has a long standing history of screaming when he wants a cigarette. Staff members usually give him a cigarette when he screams, but a behavior analyst tells staff members that it is reinforcing his behavior and they should stop giving him cigarettes. As a replacement behavior, they teach Gary to verbally ask for a cigarette but he continues screaming. This period of screaming is known as a(n) _____.
   a) extinction burst
   b) spontaneous recovery
   c) resistance to extinction
   d) response latency

21. Mario is a 7-year-old child with Autism. He does not like to hear people whistling and when he hears it, he always hits them. A behavior therapist teaches Mario to say, "Please stop whistling." The therapist whistles in front of him and when Mario says, "Please stop whistling", he immediately stops whistling; however, if Mario hits the therapist, the therapist continues whistling. What is this procedure called?
   a) DNRA
   b) DRD
   c) DNRI
   d) DISTAR

22. Mike, a fifth grade teacher decides to teach his students how to deposit cash and checks, withdraw money and get bank statements at the bank. After Mike teaches them these skills, he creates a mock bank in the classroom, prepares bank forms and uses other teachers as bank tellers. This tactic is called _____.

a) multiple exemplar training
b) programming common stimuli
c) general case analysis
d) teaching loosely

23. When shown an apple, a banana, a strawberry and a pineapple, you say, "fruits." This is an example of _____.

a) consequence stimulus class
b) antecedent stimulus class
c) feature stimulus class
d) arbitrary stimulus class

24. You are a BCBA working with adults with developmental disabilities. You have a research project. You want to involve your clients in your research and give them money for their participation. In order to do so which of the following do you need to consider?

a) You must contact their parents and get an approval from them.
b) You must get an approval from the Institutional Review Board or the Human Rights Committee.
c) You must recognize their participation, effort, and include their names in the publications.
d) If your client agrees, you don't need to get any approval from others.

25. Victoria is taking a singing lesson at home. Her mother tells her school teacher that her singing skills have improved, so her teacher asks Victoria to sing a song in front of her students. However, Victoria is shy and does not want to sing in front of them. This phenomenon is known as a(n) _____.

a) overgeneralization
b) transitive CMO
c) masking
d) negative punishment

26. Every time Jacob hears his neighbor playing the drums, he closes his window so that his room is quiet. This is an example of a _____.

a) positive reinforcement
b) negative reinforcement
c) positive punishment
d) negative punishment

27. Which of the following is known as ordinate?

a) Baseline
b) X axis
c) Condition change line
d) Y axis

28. Which of the following is true about exclusion time-out?

a) It is always used as a form of negative reinforcement.
b) It teaches appropriate behaviors.
c) When time-out ends, even if one continues misbehaving, time-out should still be over.
d) Even if one is sent to a time-out room, the door should be unlocked.

29. Which of the following is known as ripple effect, spillover effect and vicarious reinforcement?

a) Overcorrection
b) Generalization across subjects
c) Response generalization
d) Independent group contingency

30. Harold smokes approximately 15 cigarettes during waking hours from 7:00am to 10:00pm. When he craves a cigarette, he starts engaging in property destruction. You decide to implement NCR with the cigarettes. What would be the ideal implementation of NCR for this case?

a) Fixed interval schedule of every hour during waking hours
b) 10 cigarettes a day
c) 14 cigarettes a day
d) Free access to cigarettes anytime Harold wants them

31. Which of the following method uses Standard Celeration Chart?

a) Discrete trial training
b) Precision teaching
c) Pivotal response training
d) Incidental teaching

32. Erica has a 3-year-old son. When she sends him to his preschool in the morning she tells him, "Please don't play with paint today. These are very expensive clothes." What would be the best measurement system she should use for this?

a) Event recording
b) Momentary time sampling
c) Whole interval recording
d) Permanent products

33. Which of the following is an example of response cost?

a) Eric bought a new car for $20,000. He was excited for a month but he found a better car. He sold his car to get the other one but his car was only worth $12,000.
b) Linda insisted of her parents that she was going to take her favorite doll to school and put it in her backpack. Because toys are not allowed at school, her parents secretly took the doll out before she went to school.
c) Mike worked at a group home and received an $80 check. When he went to his car, he realized that he got a parking ticket for being parked during street sweeping and had to pay $50.
d) Raul had $10 in his pocket. He went to a liquor store and bought two 5-dollar lottery tickets. He did not win any money and lost $10.

34. You are tutoring a 12-year-old boy in math for two hours. He does not like math at all and he takes more than 10 bathroom breaks. You do not want him to take more than 4 bathroom breaks during tutoring. Which of the following would be an ideal procedure to use for this scenario?

a) DRH
b) DNRA
c) DRL
d) NCR

35. During a 60-minute observation, Paul engaged in hand flapping after 10, 16 and 46 minutes respectively. What would be the mean interresponse time?

a) 15 minutes
b) 18 minutes
c) 36 minutes
d) 72 minutes

36. Which of the following is NOT true about permanent product?

a) The experimenter can do other tasks while the subject is engaging in a target behavior.
b) There is less reactivity.
c) The result is considered more important than the process.
d) It often under- or overestimates the actual occurrence of a target behavior.

37. You want to compare the effectiveness of using ice cream, cotton candy and chocolate as reinforcrers for a 4-year-old boy by rapidly alternating the use of each reinforcer. This design is called _____.

a) changing criterion design
b) reversal design
c) simultaneous treatment design
d) multiple baseline design

38. All of the following is true regarding functional analysis except:

a) In the contingent attention condition, if the subject engages in the target behavior, you should immediately provide attention to him/her.
b) It includes direct observation of the subject in a natural setting.
c) Conducting a functional analysis may temporarily reinforce the target behavior.
d) It is referred to as analog.

39. Lisa speaks English. When she sees an American man and says, "How are you?" he responds to her and engages in conversation with her. When she sees a Japanese man she says, "How are you?" but he does not respond to her. Lisa then says, "How are you?" when she sees the American man but she does not when she sees the Japanese man. This is an example of _____.

a) stimulus generalization
b) response generalization
c) response discrimination
d) stimulus discrimination

40. The reversal design would be suitable for all of the following except _____.

a) self-injurious behaviors
b) riding a bicycle
c) number of hand raisings in class
d) stealing food

41. A whole interval recording tends to underestimate the overall percentage of the target behavior. If we want to have more precise data with the whole interval recording, what should we do?

a) Have a more specific operational definition for the target behavior
b) Make the interval time shorter
c) Use partial interval recording instead
d) Have more than two observers

42. You are a BCBA working with Chris, a 4-year-old boy with Autism. His parents tell you that he sometimes clenches his fists. When you observe Chris, he engages in this type of behavior on average of 2 times a week and he does it when he cannot say words correctly. What would be the most appropriate intervention you should implement as a BCBA?

a) Practicing speech so Chris would be able to say words and would not need to clench his fists
b) Giving him a stress ball so Chris can squeeze it when he needs it
c) Using DRO by giving him tokens when Chris is not clenching his fists
d) No intervention is needed.

43. When a new condition is implemented, one's behavior is still influenced by a prior condition. This is known as _____.

a) sequence effects
b) default technology
c) type I errors
d) type II errors

44. When Peter drives home after work, he realizes that his car is almost out of gas. He does not want to forget to go to the gas station so he places a $20 bill on the dashboard. The following day, when he enters his car, he sees the $20 bill and remembers to go to the gas station. This type of prompt is known as _____.

a) in-vivo contingency prompt
b) specific response prompt
c) generic response prompt
d) stimulus contingency prompt

45. Which of the following graph shows a change in level but not in trend?

a)

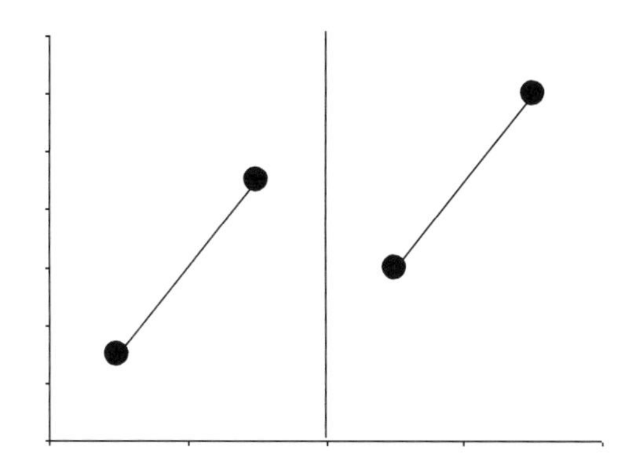

b)

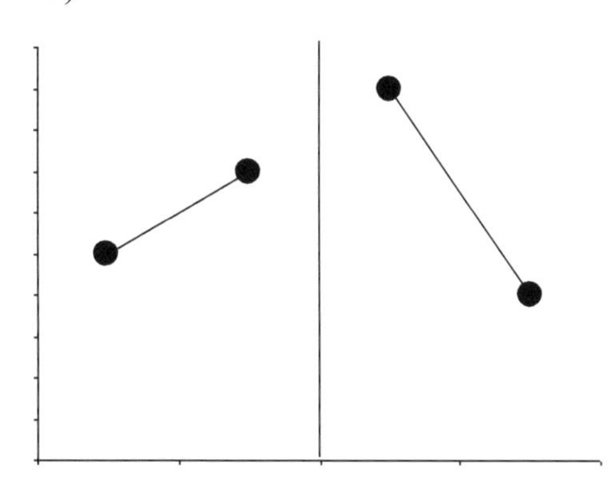

c)

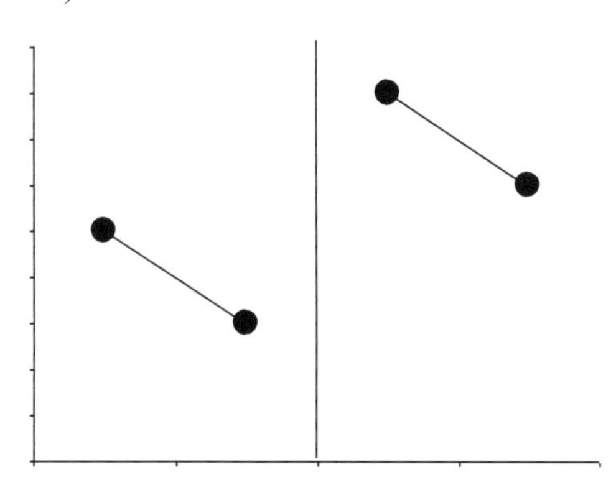

d)

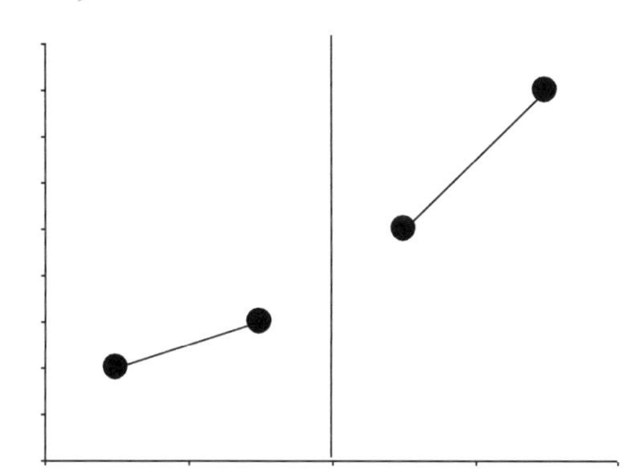

46. A _____ is also known as histogram and the summarized form is displayed to compare the effects of two or more subjects or conditions.

a) bar graph
b) semi-logarithmic chart
c) scatterplot
d) line graph

47. When an intervention is implemented, there may be other unexpected factors influencing the behavior. Those factors are also known as _____.

a) multiple treatment interference
b) unconditioned elicitors
c) artifacts
d) confounding variables

48. Mary is a two-year-old girl. When people ask her questions, she likes to repeat the questions. Her father is trying to teach her to respond to their questions instead of just repeating them. What he is trying to teach is called _____.

a) tact
b) mand
c) intraverbal
d) transcription

49. Giving an individual multiple easy tasks right before his/her non-preferred task is an intervention called _____.

a) interspersed requests
b) high-probability request sequence
c) behavioral momentum
d) all of the above

50. You conduct a functional analysis for a child's jumping behavior. Based on the results shown below, what would be the most appropriate intervention you should implement in the future?

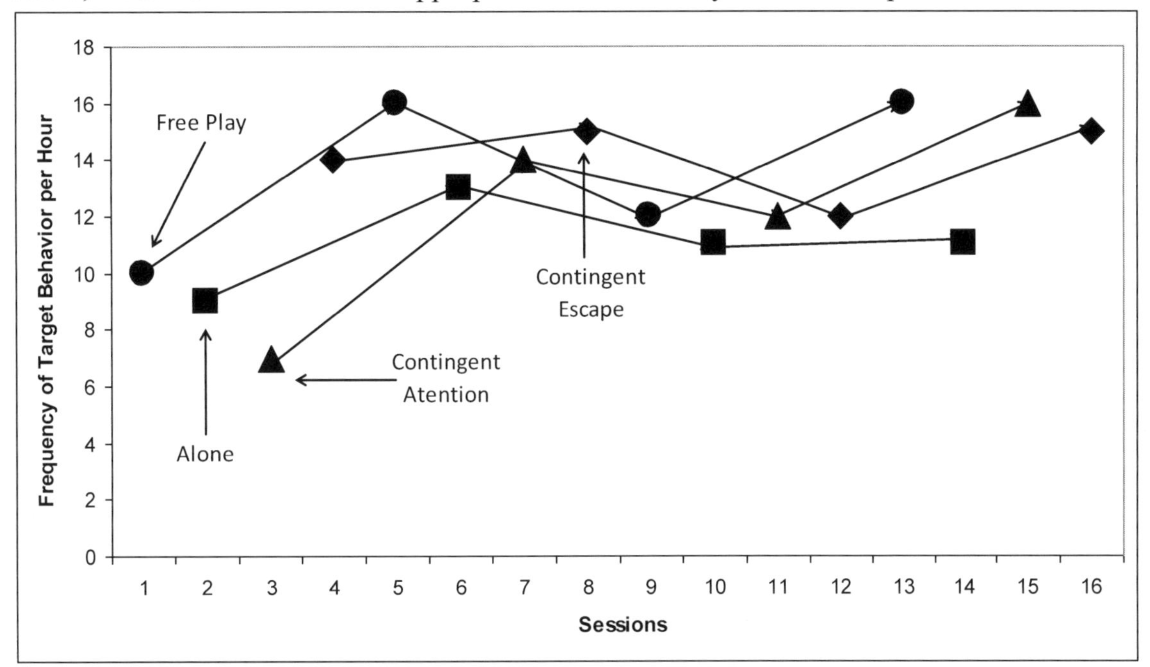


a) Teach the child how to solicit attention appropriately and ignore his jumping.
b) Teach the child how to ask for a break or stop his non-preferred activity and do not let him get away from the activity when jumping.
c) Teach him to jump on a trampoline on a fixed schedule.
d) Reprimand the child and do not let him jump.

51. You are a BCBA working with adults with Mild Intellectual Disabilities at a group home. When you have an annual progress meeting for one of them and talk about his behavioral progress with the team, he says, "I don't want you to talk about my behavior. Why does everyone need to know it? I want to skip this part." What would be the most appropriate way to respond?

a) You should say, "I know you don't want me to talk about this, but this is your annual progress meeting. We have to cover everything including your behavior."
b) You should say, "I'm sorry, but we have to discuss this. If you don't want us to talk about it, you should behave better next time."
c) You should say, "Okay, I won't talk about bad behaviors. Let me just tell them of your good behaviors and your accomplishment."
d) You should say, "I understand that you don't like this part, but it is important for the team to know how they can support you. If you don't want to hear it, would you like to step away while we are discussing this?"

52. Ken and Gabriel observed Anna's hair pulling behavior. The result is shown below.

| Interval (Time) | Ken | Gabriel |
|---|---|---|
| 1. (4:00 – 4:09) | /// | /// |
| 2. (4:10 – 4:19) | / | // |
| 3. (4:20 – 4:29) | // | // |
| 4. (4:30 – 4:39) | //// | /// |
| 5. (4:40 – 4:49) | /// | /// |

When using the exact count-per-interval IOA for this, the percentage would be _____.

a) 20%
b) 40%
c) 60%
d) 80%

53. Duration of a behavior is also known as _____.

a) latency
b) temporal extent
c) temporal locus
d) inter response time

54. Which of the following demonstrates an example of satiation?

a) Jay gets a cookie upon completing math questions. When it was on FR5 to FR10, he worked on math questions, but when it was on FR15, he stopped working.
b) Alex could not focus on watching the movie because he did not sleep well.
c) Tina likes to eat marshmallows but after eating 3 sandwiches, she does not want to eat marshmallows.
d) Cindy cleans her room when mom is around but she does not clean her room when dad is around.

55. Which of the following is NOT an example of a prompt?

a) When a child is trying to remove a tightly closed toothpaste cap, his dad loosens it a little so the child can successfully remove it.
b) After a child uses the bathroom, his mother looks at his hands and he remembers to wash his hands.
c) When a child is trying to solve an addition problem, his teacher gives him sticks to count and the child says, "3+4=8."
d) Paul always forgets to turn off the lights when he leaves home. He leaves a sticker note saying "Turn off lights" on the wall and it helps him turn off the lights when he leaves home.

56. George sometimes engages in physical aggression, self-injurious behaviors and property destruction. After conducting a functional analysis, you hypothesize that these behaviors are maintained by receiving attention from adults. You want to implement DRO with a token economy to reduce these behaviors. If you want to know whether DRO really works with these behaviors, which design would be most suited?

a) Withdrawal design
b) Multiple baseline design
c) Reversal design
d) Changing criterion design

57. You are a BCBA working in a country where corporal punishment is legally allowed and your client receives corporal punishment for not following a direction by his parents. The punishment is actually working and the client is becoming more compliant. As a BCBA, what would be your role in this situation?

a) Do nothing as corporal punishment is not illegal and it is working.
b) Immediately tell his parents to stop corporal punishment because it is against BACB guidelines.
c) Explain to his parents the disadvantages of corporal punishment and offer alternative interventions.
d) Avoid corporal punishment and teach your client to follow directions during your therapy time but do not interrupt his parents' corporal punishment during family time as it is their way of disciplining their child.

58. When there is high procedural fidelity, _____.

a) the treatment is conducted as planned
b) it has less internal validity
c) the treatment is less reliable and treatment drift may occur
d) it has a high degree of external validity

59. Tina is a mother of three children. One of her children, Kyle, spends a long time eating his meal. They all like to watch TV, so Tina tells them that once Kyle finishes eating they all can watch TV. The procedure Tina is implementing is called _____.

a) independent group contingency
b) interdependent group contingency
c) dependent group contingency
d) discrimination training

60. Which of the following is an example of a behavior cusp?

a) Learning how to hold chopsticks so one does not need to use a fork
b) Learning how to read words so one can read books, traffic signs and letters
c) Learning how to use a public phone so one can call his/her parents when away from home
d) Learning how to catch fish so one can cook and eat it

61. You are a BCBA working with Nick, a 25-year-old man with Autism. He smokes 20 cigarettes a day and often coughs. His doctor tells him that he needs to stop smoking for his health and Nick wants to quit smoking. You implement a fixed interval schedule for his smoking breaks. He agrees to follow the schedule, but one day he gets very frustrated and tells you, "I need a cigarette right now! I don't want to follow the schedule anymore!" What should you do as a BCBA?

a) Tell him that he can smoke if he wants to but inform him the health effect of smoking.
b) Tell him that he has agreed to the behavior plan and do not let him smoke a cigarette.
c) Implement a negative practice.
d) Nick is going through an extinction burst, so you should ignore his statements until he becomes calm.

62. When you watched the weather channel you found out that it would rain the following day, so you prepared an umbrella and rain boots. The following day it did not rain. Not raining serves as a(n) _____ for using the umbrella and rain boots.

a) $S^D$
b) EO
c) AO
d) consequence

63. You want to know how well your client can perform under different circumstances, so you randomly change the materials, words, tone of voice, locations and time. This procedure is known as _____.

a) multiple exemplar training
b) programming common stimuli
c) general case analysis
d) teaching loosely

64. Which of the following carefully examines two or more independent variables in the treatment conditions to determine the effectiveness and ineffectiveness of each variable?

a) Treatment analysis
b) Control analysis
c) Component analysis
d) Validity analysis

65. In which measurement does an artifact more likely occur?

a) Continuous measurement
b) Discontinuous measurement
c) Permanent product
d) Event recording

66. Andrew does not like to hear the sound of a vacuum cleaner and always hits his head when he hears it. His mother turns on the vacuum cleaner and only when he is not hitting himself, she turns off the vacuum and gives him his favorite cookie. This is an example of _____.

a) DNRA
b) DNRI
c) DNRO
d) DNRD

67. Which of the following teaching method focuses on self-initiation of the subject's preferred activity in a natural setting?
a) PRT
b) DTT
c) TEACCH
d) PSI

68. When is a differential reinforcement least effective?
a) The problem behavior cannot be ignored.
b) The schedule of reinforcement is intermittent.
c) The subject does not have the replacement behavior in his/her repertoire.
d) There are only a few available reinforcers.

69. Which of the following is NOT a verbal behavior?
a) Showing sign language
b) Showing a written word
c) Clearing throat
d) None of the above

70. Which of the following intervention functions as an AO?
a) NCR
b) Extinction
c) Premack principle
d) DRO

71. Which of the following is an example of a sensory extinction for turning a light switch on and off to see the light flickering?
a) Telling the individual to stop touching the light switch
b) Ignoring the individual to turn the light switch on and off
c) Removing the light bulb and allowing the individual to turn the light switch on and off
d) Blocking the individual from touching the light switch

72. A post-reinforcement pause tends to be observed in _____.
a) FI and VI
b) FI and FR
c) FR and VR
d) FR and VI

73. Dianna is a very shy girl. When she sees someone new, she speaks very softly and they cannot hear what she is saying. In order for others to hear what she is saying, _____ must be changed.
a) topography
b) celeration
c) magnitude
d) rate

 

74. Eugene decided to exercise every day. He chose to run to a yogurt store and get one cup of yogurt as a reinforcer for running. One day, he was very tired after work but wanted to have a cup of yogurt. Instead of running, he rode a bicycle and said to himself, "Well, bicycling is still an exercise" and got himself yogurt. In applied behavior analysis, this is called a _____.

a) contingency contrast
b) bootleg reinforcement
c) faulty self-management
d) measurement bias

75. Lucy, a BCBA works in a laboratory using mice. She wants to try a new experiment using a reversal design and asks her co-worker, Dan, to collect data. Dan tells her that he can observe it for a short time but has no time to collect data. Lucy then hires an untrained staff person, Bill, to collect data. Lucy explains to Bill the purpose of this experiment before he collects data, but because he is not used to collecting data, Lucy finds out that it is not accurate. She then hires an experienced staff person, Charles, to collect data. Lucy does not tell him the purpose of her experiment but explains to him how to collect data. Lucy gathers accurate data from Charles. During this experiment, Lucy's boyfriend, Gary visits her lab and observes how people are conducting the experiment. In this scenario, who is the naive observer?

a) Dan
b) Bill
c) Charles
d) Gary

76. Which of the following is NOT an antecedent intervention?

a) High-probability request sequence
b) Extinction
c) NCR
d) Priming

77. Nicholas is a toddler. His mother praised him when he crawled. She then praised him only when he stood. Now, she praises him only when he walks. What is this teaching technique called?

a) Chaining
b) Total task
c) Prompt fading
d) Shaping

78. An intermittent schedule of reinforcement is used when _____.

a) a new behavior needs to be taught
b) a behavior has been learned and needs to be maintained
c) a subject needs to maintain a momentum
d) an acquisition rate is becoming low

79. Because you always heard an unpleasant squeaking sound when you opened a bathroom door, you got goose bumps just by touching the door knob. One day, you put oil on the hinge to fix it. Because you did not hear the squeaking sound anymore, you gradually stopped getting goose bumps when you touched the door knob. This is an example of _____.

a) respondent extinction
b) reflex transformation
c) conditioned stimulus fading
d) response blocking

80. James is a 3-year-old boy with Autism. During ABA therapy, a therapist asked him to clap his hands. When he clapped his hands, the therapist said, "Good job!" and sang Itsy Bitsy Spider. James hates spiders and started crying. Since then he stopped clapping his hands when asked to clap. This is an example of _____.

a) positive reinforcement
b) negative reinforcement
c) positive punishment
d) negative punishment

81. Ryan is receiving ABA services. Because he often throws a tantrum when he sees non-preferred food on the table, his therapists decide to work on reducing his tantrums. Which of the following would be the most appropriate behavioral goal for this?

a) Ryan will stop throwing a tantrum for five consecutive days.
b) Ryan will eat his non-preferred food for three consecutive days.
c) Ryan will be reinforced with his favorite toys when he does not throw a tantrum upon seeing non-preferred food.
d) Ryan will engage in a replacement behavior instead of throwing a tantrum, 4 out of 5 opportunities, upon seeing non-preferred food.

82. Which of the following is NOT an unconditioned reinforcer?

a) Verbal praise
b) Sleep
c) Air
d) warmth

83. Which of the following would be an example of indirect measurement?

a) Surveys
b) Whole interval recording
c) Event recording
d) Permanent product

84. Which of the following is an example of social negative reinforcement?

a) Catherine does not like Bruce. She knows Bruce goes to the gym on Friday, so she avoids going to the gym on Friday.
b) Peter jokes in class. His teacher does not like it and often reprimands Peter but because his classmates always laugh after he jokes, he continues joking.
c) When Gail is doing her homework, the volume on the TV that her brother is watching is too loud. She asks him to turn the volume down. Her brother turns the volume down so she can do her homework.
d) Liz does not like to eat pizza. However, when she is at school, she eats it because she does not want her friends to make fun of her.

85. William hits his classmate during PE class. His PE teacher tells him to do 50 pushups for hitting his classmate. This procedure is called _____.

a) response blocking
b) contingent exercise
c) positive practice
d) negative practice

86. You are a BCBA working with David, a 16-year-old man with Mild Intellectual Disability and Schizophrenia. His behaviors have improved since you implemented behavioral strategies. However, he sometimes talks to the wall and engages in self-injurious behaviors. His parents want you to implement other behavioral strategies to reduce his self-injurious behaviors. What should you do?

a) You should conduct a functional assessment and determine the function of David's self-injurious behaviors.
b) You should continue implementing the same behavioral strategies as his self-injurious behaviors are occurring due to extinction burst.
c) You should advise his parents to talk to his doctor as it could be occurring due to schizophrenia.
d) You should tell his parents to give him attention non-contingently because his self-injurious behaviors are occurring due to lack of attention.

87. Alvin wanted to know how fast his client, Tom, could write 50 words and asked Natalie to record the data. Natalie had Tom write 50 words every day but instead of recording the duration, she carefully recorded the number of words Tom spelled correctly. Based on this scenario, Natalie's data is NOT _____.

a) valid
b) effective
c) accurate
d) reliable

88. Which of the following is NOT a level of scientific understanding?

a) control
b) replication
c) prediction
d) description

89. Derek did not spend enough time brushing his teeth. His parents bought a sand timer and told him that he had to keep brushing his teeth until the top glass of the sand timer is empty. When Derek succeeded, his parents gave him a star on a token board and when he received five stars, his parents bought him a baseball. In this scenario, which of the following is a dependent variable?

a) Duration of tooth brushing
b) Sand timer
c) Star
d) Baseball

90. You are a BCBA, working with Michelle, a 22-year-old woman with Autism. Michelle's brother asks you to work on her compliance. Her compliance has never been an issue but her brother wants her to do his favors such as turning on the TV for him or getting a newspaper for him. If you decided to work on these strategies, which dimension of ABA would you be violating?

a) Analytic
b) Technological
c) Behavioral
d) Applied

91. In which of the following scenarios is PLACHECK most useful?

a) Basketball game score
b) Cleanliness of dishes
c) Engagement of math assignment
d) Duration of singing

92. Frank is a doctor who is performing a very important heart surgery in two days. Frank practices the surgery by watching a video of a famous heart surgeon and imitating the hand motions in the air. The behaviors by the famous surgeon and Frank have _____.

a) functional relation
b) formal similarity
c) transitivity
d) reflexivity

93. Which of the following exemplifies a behavior trap?

a) Ed told Sandy that she would get $5 if she cleaned his car. After Sandy cleaned the car, Ed drove away without giving her $5. Sandy stopped cleaning Ed's car.
b) When Elisa was eating salad at Saladicious, she found a bug. Since then, she has not eaten a salad at Saladicious.
c) Greg wants to play tag with his peers but does not know how to initiate it. His mother teaches him to say, "Can I play?" to his peers. Greg says, "Can I play?" and joins his peers. He now says, "Can I play?" when he sees his peers.
d) Cody did not like to take his medicine. Cody's mother added his medicine to his tea and he drank it without knowing it. Cody's mother continued putting his medicine in his tea.

94. Which of the following is not a part of an indirect functional assessment?
a) Scatterplot
b) rating scale
c) checklist
d) interview

95. When Ellen uses the washing machine, she often forgets to pour soap or gets confused with pressing the buttons in order. Which teaching method should be used to teach her how to use the washing machine correctly?
a) Shaping
b) Premack Principle
c) Chaining
d) Token Economy

96. Melissa engages in scratching her arm. After conducting a functional analysis, it is determined that the behavior is maintained by automatic reinforcement. If her behavior analyst decides to place this behavior on extinction, what do people need to do?
a) They should ignore Melissa's arm scratching.
b) They should physically block Melissa's arm scratching.
c) They should redirect Melissa to rub her arm instead of scratching.
d) They should apply a towel or long sleeve on Melissa's arm.

97. Victor's car stereo broke and he could not get his CD out of it. It has been a couple of months now and even though Victor knows it is broken, he sometimes presses the eject button in hopes that his CD would come out. This behavior is known as a(n) _____.
a) extinction burst
b) conditional probability
c) placebo control
d) spontaneous recovery

98. Two observers recorded the occurrence and non-occurrence of a target behavior and the data is shown below.

| Interval # → | 1 | 2 | 3 | 4 | 5 | 6 | 7 | 8 | 9 | 10 |
|---|---|---|---|---|---|---|---|---|---|---|
| Observer 1 | X | X | X | X | O | O | X | O | X | O |
| Observer 2 | X | O | X | X | O | O | X | X | X | X |

X: Occurrence of the target behavior O: Non-occurrence of the target behavior

What would be the percentage of unscored-interval IOA?
a) 40%
b) 50%
c) 60%
d) 80%

 

99. Ken is a program supervisor for an Autism program. His client, Dave, had a lot of self-injurious behaviors, so he designed a behavior plan using DRO and had his therapists implement it during their sessions. Three months later Ken collected Dave's behavior data and found out that the frequency had not changed at all compared to the baseline. Ken thought that his behavior plan was not working; however, in fact, the severity of Dave's self-injurious behaviors had decreased. Because the severity level was not recorded, Ken did not notice Dave's progress. This is an example of a(n) _____.

a) measurement bias
b) false negative
c) artifact
d) false positive

100. Which of the following is NOT true regarding private events?

a) Private events are overt behaviors.
b) Private events include thoughts and feelings.
c) Private events are described in radical behaviorism.
d) B.F. Skinner believed private events are also behaviors.

101. George owns a factory with 50 employees. He is usually in the office but he wants to evaluate his employees' performance and walks around the factory to observe them. As his employees feel closely monitored, they start working harder than usual. George is not able to observe their natural performance due to a(n) _____.

a) observer drift
b) abative effect
c) unconditioned response
d) reactivity

102. Stacy is a BCBA working with Douglas, a 27-year-old man with Moderate Intellectual Disability at a day program. His service coordinator at the regional center tells Douglas that she wants to transfer him to a paid work program when he shows appropriate behaviors and exhibits verbal aggression fewer than two times per month. Stacy is very busy, so she forgets to collect his behavior data. She thinks that Douglas is doing very well, so she tells his service coordinator that Douglas should be transferred to the paid work program. Which attitude of science is Stacy's subjective statement violating?

a) Determinism
b) Empiricism
c) Parsimony
d) Experimentation

103. When Gloria teaches a child how to brush his teeth, she gives him a 3-second fixed time delay. This time delay exists between _____.

a) the presentation of the $S^D$ and the prompt
b) the presentation of the $S^D$ and the response
c) the response and the delivery of a reinforcer
d) the response and the next $S^D$

104. Peter is a fitness trainer for children. He tells his students that if they do 10 sit-ups they will get a sticker. They all start doing 10 sit-ups. Peter realizes that a sticker is a strong reinforcer. Peter then gradually increases the number of sit-ups. One day, when he tells them to do 50 sit-ups for a sticker, they stop doing sit-ups. When he asks them if they do not want the stickers anymore, they say they still do. What could be the most plausible factor for their discontinuation?

a) Satiation
b) Extinction burst
c) Ratio strain
d) Post-reinforcement pause

105. What is wrong with the following cumulative graph?

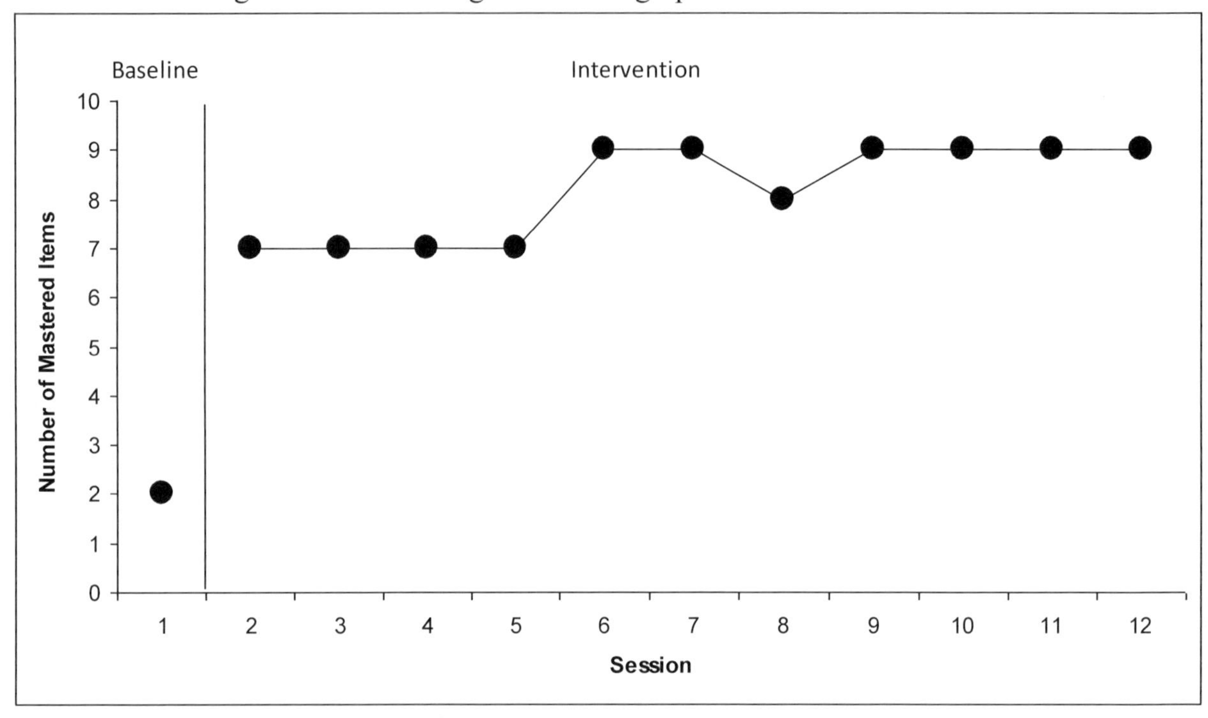


a) The baseline period is too short.
b) The progress is made too quickly in the first intervention session.
c) The data descends.
d) There is no trend line.

106. Kim is a BCaBA who works with children with Autism. Her uncle, Richard, tells Kim that his friend's son was recently diagnosed with Autism and the parents want Kim to conduct a Functional Behavior Assessment. Because Kim has been working at an ABA agency for a long time, she is familiar with an FBA and she is competent with conducting it. Which of the following would be the most appropriate way for Kim to conduct an FBA?

a) Kim knows how to conduct an FBA and is competent; therefore, she should start conducting it.
b) Kim does not have a BCBA; therefore, she cannot conduct an FBA.
c) Kim needs to talk to a BCBA and conduct an FBA under the BCBA's supervision.
d) Since it is referred by Kim's uncle, she cannot conduct an FBA due to a conflict of interest.

107. Mike is a father of a five-year-old boy. He wants to increase his son's compliance. His friend who has a BCBA teaches him various ABA methods to increase his compliance but Mike pays little attention to them and after trying several methods for a very short time, he concludes that spanking and yelling are the most effective ways. This exemplifies _____.

a) stimulus salience
b) arbitrary stimulus control
c) false claims
d) default technologies

108. Jack is an 11-year-old boy with Autism. Because he does not have money skills, his therapists taught him how to purchase items using fake money. Jack learns very quickly and gives the right amount of money while engaging in a pretend shopping event. However, when Jack goes to a real store with real money, he does not know what to do. Jack lacks _____.

a) social validity
b) external validity
c) treatment validity
d) internal validity

109. Which of the following best describes a backup reinforcer?

a) It is a reinforcer that can be given when a primary reinforcer is not available.
b) It is a reinforcer that can be purchased with earned tokens.
c) It is a reinforcer that can be used when a subject does not meet the criterion.
d) It is a reinforcer that can be given when a subject meets an additional goal.

110. Which of the following would be an example of a contingent observation?

a) Tina agrees to observe Tony's physical aggression. She only observes and records his physical aggression but not other types of behaviors.
b) Omar observes a therapist and confirms that an intervention is correctly implemented.
c) Gus receives a time-out while playing baseball. He sits on the bench and watches other people playing baseball.
d) Chris talks to the teacher without raising his hand in class. His teacher ignores him and picks other students who raise their hands.

111. Henry wants his younger brother to clean up their room and if his brother says no, Henry tickles him, so his brother eventually cleans it up. Every time Henry tickles his brother, his brother giggles. His brother's giggling is a _____.

a) respondent behavior
b) operant behavior
c) conditioned stimulus
d) negative reinforcement

112. Jose works at a food bank. When he yells, "I need a cigarette break now!" to his job coach, Mary, she gets scared and gives him a cigarette break. When he yells to another job coach, Robert, he tells Jose, "It's rude to yell at a person" and ignores his request. Jose then yells for a cigarette break only when Mary is around. This phenomenon is called _____.

a) stimulus generalization
b) stimulus discrimination
c) response generalization
d) response discrimination

113. Kate's roommate tells her that she is from Brazil. Kate later learns that people from Brazil speak Portuguese. Kate then says, "My roommate must be able to speak Portuguese." This is an example of _____.

a) transitivity
b) paring
c) symmetry
d) reflexivity

114. Which of the following is known as abscissa?

a) Baseline
b) X axis
c) Condition change line
d) Y axis

115. What would be the response latency for putting shoes on?

a) Time between the mother saying, "put your shoes on" and the child initiating putting his shoes on
b) Duration of the child putting his shoes on
c) Time between the mother saying, "put your shoes on" and the child completing putting his shoes on
d) Time between the child completing putting his shoes on and the child receiving a candy

116. Mark is an eight-year-old boy who likes to play videogames. His mother does not allow him to play videogames on Saturday. On Sunday, she tells Mark that he can play videogames only if he mows the lawn. Mark's mother is manipulating his _____.

a) $S^D$ control
b) AO
c) antecedent
d) EO

117. Multiple baseline design can be conducted across all of the following except _____.

a) subjects
b) treatments
c) settings
d) behaviors

118. Ray believes that an energy drink, Xtreme Power 1000, keeps people awake longer. He gathers two participants and has one drink Xtreme Power 1000, and another drink sugar water, to see how long they can stay awake over 10 sessions. To eliminate a biased perception, Ray covers the labels on both drinks, so neither the participants nor Ray know whether it is an energy drink or sugar water until the research is over. This procedure is called _____.

a) placebo experiment
b) naive observation control
c) arbitrary stimulus experiment
d) double-blind control

119. A toddler gets a shot from a doctor. When the toddler sees the doctor, he starts to cry. The doctor is a _____.

a) conditioned stimulus
b) conditioned response
c) unconditioned stimulus
d) unconditioned response

120. Which of the following has a point-to-point correspondence?

a) When seeing a bicycle, you say "bicycle."
b) When hearing "guess what?" you say "what?"
c) When seeing "dinosaur" written, you say "dinosaur."
d) When seeing a red light, you say "stop."

121. A train arrives at Glen Oaks Station every 20 minutes. When the train arrives at the station, the door remains open for 15 seconds and the train leaves. This demonstrates _____.

a) FI 20-minute schedule with spaced responding 15-second
b) VI 20-minute schedule with limited hold 15-second
c) FI 20-minute schedule with limited hold 15-second
d) VI 20-minute schedule with spaced responding 15-second

122. When there is a high degree of variability, _____.

a) there is a high response rate
b) an independent variable is showing little or no effect
c) there is a high degree of a trend
d) an intervention should be introduced immediately

123. Javier's father tells him to be careful with bees because they can sting him. When Javier sees a bee, he runs away so he won't get stung by the bee. This is an example of _____.

a) rule-governed behavior
b) high-probability request sequence
c) contingency-shaped behavior
d) situational inducement

124. Martha is in the seventh grade. She likes Tim and one day Tim sits next to her. While she is staring at him, the history teacher asks her a question. Martha cannot answer it. This effect is called _____.

a) reactivity
b) overshadowing
c) ripple effect
d) indiscriminable contingency

125. Which of the following is an example of imitation?

a) Brian says "jump" and Ben jumps.
b) Dean sings the national anthem on Thursday and Kevin sings the national anthem on Friday.
c) At 5:00 pm, many people take the 10 Freeway and head East.
d) Dorothy watches the fitness video and copies the instructor's action.

126. Anthony belongs to a tennis club. After his tennis lesson, if he does not clean the tennis balls, his trainer has him do 100 swings. Because Anthony does not like to do 100 swings, he cleans the tennis balls. This is an example of _____.

a) positive reinforcement
b) negative reinforcement
c) positive punishment
d) negative punishment

127. When all students in the class submitted their art assignments, they all received tickets to the museum. This procedure is called a(n) _____.

a) dependent group contingency
b) hero procedure
c) independent group contingency
d) interdependent group contingency

128. Which of the following is NOT a conditioned response?

a) Kelly opening the door after hearing a doorbell
b) Gina asking for water when seeing her friend drinking water
c) Greg sneezing after breathing pepper in his nose
d) Steve looking at his watch when asked what time it is

129. Nina wants her son to put on his pajamas at bedtime. When he wears his pajamas, Nina reads a book to him at bedtime. When he complains and does not wear his pajamas, he has to sleep without a bedtime story. Her son starts putting on his pajamas prior to bedtime. Which of the following is an independent variable?

a) Wearing pajamas independently
b) A bedtime story
c) Complaint
d) Sleeping without a bedtime story

130. Which of the following is an example of an explanatory fiction?

a) Ted always fixes the picture frames on the wall because he has Obsessive Compulsive Disorder.
b) Mary is screaming because she has not eaten food for more than 5 hours.
c) Keith is a good football player because he practices football every day.
d) Erwin stopped talking when his teacher walked into his classroom.

131. Peter works at a zoo. When he was cleaning a bear's cage, he got scratched by a bear. After that incident, whenever he sees the bear's cage his body shivers. The bear's cage is a(n) _____.

a) US
b) CR
c) UR
d) CS

132. Temporal locus is _____.

a) when the target behavior occurs
b) how long the target behavior lasts
c) when the target behavior ends
d) where the target behavior occurs

133. After Joseph runs 15 miles, he gets thirsty. He sees a vending machine and buys water. As he drinks the water, he is no longer thirsty. Which of the following shows a four-term contingency based on this scenario?

a) Running 15 miles – Being thirsty – A vending machine – Buying water
b) Being thirsty – Seeing a vending machine – Buying water – Drinking water
c) Being thirsty – Buying water – Drinking water – Quenching thirst
d) Being thirsty – A vending machine – Buying and drinking water – Quenching thirst

134. Which of the following would be an ideal fading order?

a) VR – FR - CRF
b) CRF – FR – VR
c) FR – CRF – VR
d) CRF – VR – FR

135. Which of the following is an example of a contingency independent antecedent event?

a) Seeing a new animal and running away
b) Preparing breakfast at night so one does not have to wake up early
c) Skipping lunch and having a long tantrum before dinner time
d) Seeing a stop sign and stopping the car

136. Elisa often bit her nails when she was bored. Her mother bought a portable videogame for her and when she looked bored, her mother told her to play the videogame. The videogame gave Elisa an automatic reinforcement and she stopped biting her nails. This would be an example of _____.

a) DRL
b) DRI
c) DRO
d) DRH

137. John is a toddler. He likes to talk to his mother on the phone. When he hears a real phone ringing and answers it, he can talk to his mother. When he answers a toy phone, he cannot talk to his mother. The toy phone is a(n) _____.

a) AO
b) $S^P$
c) S-Delta
d) punisher

138. All of the following requires the observer to directly observe the target behavior except _____.

a) duration
b) response latency
c) trials to criterion
d) frequency

139. Standard Celeration Charts have a doubling celeration angle of _____ degree.

a) 12
b) 24
c) 34
d) 48

140. When using a momentary time sampling, the observer checks whether a target behavior is occurring _____.

a) at the beginning of each interval
b) in the middle of each interval
c) at the end of each interval
d) at randomly selected time of each interval

141. It is ideal to use a changing criterion design for all of the following situations except:

a) Increasing the number of new words a student can read
b) Decreasing the number of cigarettes a man smokes
c) Increasing days to attend school
d) Increasing the number of math questions a student can solve per minute

142. In the diagram shown below, If A represents an $S^D$ and B represents a response, C would be a(n) _____ and D would be a(n) _____.

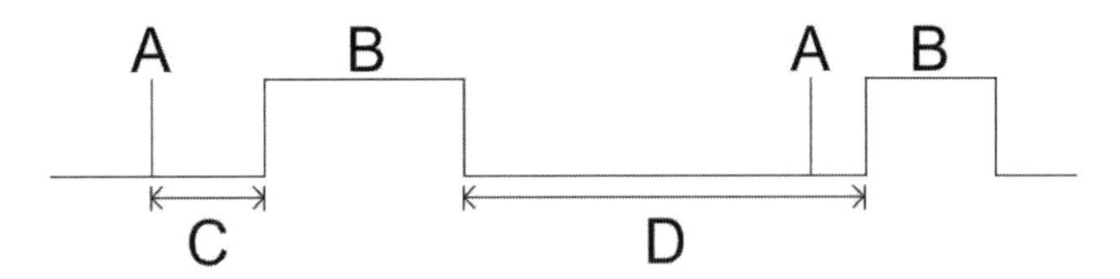


a) interresponse time, response latency
b) duration, interresponse time
c) response latency, duration
d) response latency, interresponse time

143. Moe and Gloria counted the number of Harry's self-injurious behaviors and recorded data per 30-second intervals. The results are shown below. What would be the percentage if they use an exact count-per-interval IOA?

| Interval # → | 1 | 2 | 3 | 4 | 5 | 6 | 7 | 8 | 9 | 10 |
|---|---|---|---|---|---|---|---|---|---|---|
| Moe | //// | /// | /// | // | //// | / | /// | // | 0 | /// |
| Gloria | // | /// | /// | // | //// | 0 | /// | /// | 0 | // |

a) 40%
b) 50%
c) 60%
d) 70%

144. Which of the following is NOT true about a punishment procedure?

a) Punishment does not teach appropriate behaviors.
b) The intensity of punishment should start low and gradually be increased.
c) If giving a big hug decreases the frequency of saying, "Hi", the big hug is considered punishment.
d) Implementing a punishment procedure may evoke aggressive behaviors.

145. After Sara teaches Nathan how to make a bed using a chaining procedure, she purposely sits on the bed while he is making it and waits for him to say, “Can you move?” This procedure is known as _____.

a) forward chaining with leap aheads
b) BCIS
c) PLACHECK
d) reactivity assessment

146. Permanent product focuses on _____.

a) frequency
b) result
c) duration
d) topography

147. Which of the following is NOT recommended when providing training feedback?

a) Feedback should always include positive comments.
b) Feedback should be written down.
c) Feedback should be provided after the training course is over.
d) Feedback should include role playing.

148. Aaron likes to cook and please people. When he adds 2 teaspoons of vinegar for sweet and sour soup, people say the soup is too weak. When he adds 4 teaspoons of vinegar, they say it’s too sour. Aaron then learns to add 3 teaspoons of vinegar to make good sweet and sour soup. This would be an example of a _____.

a) component analysis
b) stimulus discrimination
c) variability analysis
d) parametric analysis

149. “30+5”, “3 tens and 5 ones” and “thirty five” are an example of a(n) _____.

a) stimulus equivalence
b) arbitrary stimulus class
c) stimulus salience
d) transfer of stimulus control

150. Douglas, a BCBA, interviews Jake’s parents. They tell Douglas that Jake often engages in bad behaviors. Douglas asks them why Jake has bad behaviors but they do not know. Douglas then asks them to remember the antecedents of his bad behaviors in the past and determines that those behaviors are escape-maintained. He then discusses a behavior plan. Later, Douglas continues interviewing them to make sure that the behavior plan is working. Douglas is violating the assumption of _____.

a) determinism
b) empiricism
c) parsimony
d) experimentation

151. Which of the following shows a post-reinforcement pause with a high, steady rate?
  a) Fixed ratio
  b) Fixed interval
  c) Variable ratio
  d) Variable interval

152. Sandra bites her arm when people come close to her. When they leave, she stops biting. She also hits her head when her feet are wet. When they are dry, she stops hitting them. She pulls her hair when she hears a door bell. When the door bell stops, she stops pulling her hair. What function are these behaviors hypothesized to be?
  a) Attention-maintained
  b) Access to tangible-maintained
  c) Sensory-maintained
  d) Escape-maintained

153. George is a 6-year-old child with Autism. When he hits his classmate at school, the principal reprimands him, calls his mother and has her pick him up. George likes to see his mother, so when he wants to see her, he hits his classmate. George's behavior is _____.
  a) punished by the antecedent
  b) punished by the consequence
  c) reinforced by the antecedent
  d) reinforced by the consequence

154. Which of the following describes hero procedure?
  a) When you implement a behavior plan, it shows significant improvement.
  b) An intervention implemented only involves positive reinforcement.
  c) Only the first person who meets the preset criterion would get a reward.
  d) You choose an individual or a small group of people and when they achieve the preset criterion, everyone gets a reward.

155. Why is using a reversal design sometimes considered unethical?
  a) Because it is time-consuming.
  b) Because it reverses an effective intervention and new problem behaviors may arise.
  c) Because it reinforces the problem behavior while it extinguishes other appropriate behaviors.
  d) Because it goes back to the baseline condition to prove that the IV is effective and the problem behavior would continue occurring in the baseline condition.

156. A _____ is also known as a frequency polygon and it is the most commonly used Cartesian plane based format in ABA.
  a) Scatterplot
  b) Bar graph
  c) Line graph
  d) Semi-logarithmic chart

157. When Adam comes home, he has to press the security codes to unlock the door. Once he starts pressing the codes, he has 10 seconds to complete the codes and turn the knob or it will lock again. This contingency is known as _____.

a) concurrent schedule of reinforcement
b) behavior chain with a limited hold
c) behavioral contrast
d) spaced responding DRA

158. Which of the following is NOT a required element in baseline logic?

a) Replication
b) Observation
c) Prediction
d) Verification

159. Hilda is an experienced therapist who works with a child with a severe self-injurious behavior. The behavior is maintained by automatic reinforcement. When this child engages in a self-injurious behavior, Hilda puts her arm around the child's body and prevents him from hitting his body. This way, his automatic reinforcement is not delivered. Hilda's intervention is known as _____.

a) masking
b) extinction
c) positive practice
d) response blocking

160. Which of the following is NOT an example of a punishment procedure?

a) When Isaiah finishes washing the dishes, his mother kisses him in front of his friends. Isaiah stops washing the dishes.
b) When Dixie pours coffee in a paper cup and grabs it, she burns her fingers. She stops using a paper cup for coffee.
c) Monica's father heard her friends shoplifted. Because he does not want Monica to shoplift, he tells her that he would spank her if she shoplifted and Monica never shoplifts.
d) Scott burps in his class and his classmates laugh at him. He laughs with them too, but he stops burping in his class.

Answers

1. c
Since it is operant, the behavior must be controlled by the history of consequences. If the behavior is induced by the antecedent, it is respondent. In an operant extinction, those operant behaviors (calling her friends on her phone) no longer produce reinforcement (being able to talk to her friends on her phone.)

2. a
The teacher is an antecedent stimulus. Yelling is a punisher as it is the consequence of talking in class.

3. c
Signing and saying has no point-to-point correspondence even though they both mean "hello".

4. a
It is an EO because it makes the reinforcer valuable. If the house was not on fire, evacuating would not have any value.

5. b
A within stimulus prompt is a cue added/modified in its physical form to stand out and lead to a correct response.

6. c
Direct Instruction is a fast-paced group teaching method developed by Siegfried Engelmann.

7. d
In a negative practice, when a child engages in an inappropriate behavior, he/she must perform the same inappropriate behavior repeatedly until he/she experiences fatigue.

8. c
Determinism states that every behavior is lawful and it is caused by a reason. A behavior does not just happen out of blue.

9. a
In a generalization phase, the client should not need any prompts or continuous reinforcement. The schedule of reinforcement should be decreased to the natural rate; therefore, it is an indiscriminable contingency.

10. b
Make sure that the behavioral statement is observable. Emotion is not observable.

11. d
Antecedent intervention (blocking access to the marbles) is necessary to prevent this life threatening behavior before other replacement behaviors are taught.

12. c
Inter-response time is the time between two responses.

13. b
PSI (Personalized System of Instruction) is a self-paced teaching method that was developed by Fred S. Keller.

14. a
It is important to keep the schedule as stated in the behavior contract. If Eric often forgets to drink, you may need to talk to him, revise the behavior contract, and reduce the frequency of drinking.

15. c
If applying NCR, the interval to deliver the reinforcer should be a little less than 15 minutes.

16. d
Symmetry is a type of stimulus equivalence that shows stimulus relations of "if A=B, then, B=A"

17. d
Three branches of behavior analysis are ABA, Behaviorism and EAB.

18. a
A line graph displays the variability of behaviors.

19. c
A parametric analysis studies the effects of a behavior by changing the value of the independent variable.

20. c
If Gary's behavior increases, it could be an extinction burst; however, the question states that he continues screaming; therefore, it is a resistance to extinction.

21. a
DNRA (Differential Negative Reinforcement of Alternative Behavior) is a procedure that uses negative reinforcers. Mario's verbalization is reinforced by withdrawing the aversive stimulus (whistling).

22. b
Programming common stimuli is used to generalize a behavior by creating the settings similar to the actual ones.

23. d
An arbitrary stimulus class evokes the same response ("fruits") when two or more stimuli that do not share a common physical form are present (an apple, a banana, a strawberry and a pineapple.)

24. b
An approval from the Institutional Review Board or the Human Rights Committee is needed for this. See BACB guidelines for responsible conduct for behavior analysts 10.15

25. c
In masking, a behavior may be maintained by a stimulus control; however, the evocative function can be blocked when another stimulus is present.

26. b
Jacob's behavior (closing the window) increases; therefore, it is "reinforcement". The drumming sound is removed as a consequence of his behavior; therefore, it is "negative".

27. d
Y axis is known as ordinate or vertical axis.

28. d
The door should always be unlocked for safety.

29. b
Ripple effect, spillover effect and vicarious reinforcement are all the same as generalization across subjects.

30. a
Harold is awake for 15 hours a day (from 7:00am to 10:00pm) and he smokes 15 cigarettes a day. He smokes on average of 1 cigarette per hour. In NCR, the rate of providing the reinforcer should meet the current rate of smoking; therefore, it should be given every hour.

31. b
Precision teaching was developed by Ogden Lindsley and it uses a semi-logarithmic chart, which is also known as a Standard Celeration Chart.

32. d
In permanent products, a behavior is measured by checking the results in the environment. When Erica's son has paint on his clothes, she would know that he has played with paint.

33. c
Response cost is a punishment procedure which involves the loss of an earned reinforcer. A fine is a form of response cost.

34. c
In DRL, the child is reinforced only when he engages in the behavior (taking a bathroom break) less often than a preset criterion. For example, if the child only takes 3 bathroom breaks during math, he can play videogames for 20 minutes after math.

35. b
There are two interresponse times in this scenario (between 10-16 and between 16-46) and each interresponse time is 6 minutes and 30 minutes. The mean interresponse time is calculated by (6+30)/2 and the answer is 18.

36. d
Permanent product focuses on the result so the observer does not have to be present while the target behavior is occurring. The measurement methods that under- or overestimates the actual occurrence of a target behavior are discontinuous measurements such as whole-interval recording, partial-interval recording and momentary time sampling.

37. c
Simultaneous treatment design is implemented by rapidly alternating the use of 2 or more reinforcers to find out the effectiveness of the reinforcers. It is also called alternating treatment design.

38. b
When conducting a functional analysis, you are testing the occurrence of the target behaviors in a contrived setting.

39. d
In stimulus discrimination, a subject learns that a behavior is reinforced in one stimulus condition but not reinforced in another.

40. b
Once one learns how to ride a bicycle, he/she cannot unlearn it, so the reversal design would not be suitable for this.

41. b
When the interval time is shortened, there would be more intervals and the percentage would become more precise.

42. d
The frequency of this behavior is low (twice a week) and it is typical; therefore, interventions would not be necessary for this. You should focus on other inappropriate behaviors instead.

43. a
Sequence effects can be observed when there is a condition change. Because the subject is accustomed to behave under the previous condition, the behavior would not change immediately after the condition change.

44. c
A generic response prompt is a cue with a general item such as a rubber band, tape, paper, etc. that leads to engage in a target behavior.

45. c
When comparing two conditions, if the data points overlap in Y-axis, there is no change in level. If the data paths show the same angle, there is no change in trend.

46. a
A bar graph uses different heights to compare data.

47. d
A confounding variable is an unintentional variable that changes a behavior.

48. c
In intraverbal, a verbal $S^D$ and a verbal response do not share the same point-to-point correspondence or formal similarity. Because Mary's father wants her to say something different from what she hears, it is intraverbal.

49. d
Interspersed requests, high-probability request sequence and behavioral momentum are all synonyms.

50. c
The behavior is observed in all conditions; therefore, it is hypothetically a sensory-based behavior and the child needs to be taught socially appropriate ways to engage in the same type of behavior.

51. d
We have to acknowledge the client's opinions and consider how we can discuss and support him/her with respect.

52. c
Ken and Gabriel have the same data on interval 1, 3 and 5. Because there are 5 intervals in this observation, you will divide 3 (number of intervals with the same data) by 5 (number of total intervals) and multiply by 100.

53. b
Temporal extent recognizes that all behaviors have a duration.

54. c
Satiation occurs when a reinforcer temporarily loses its effectiveness after satisfying the needs and wants.

55. c
A prompt is a stimulus that assists the subject to evoke a correct response. If the response is incorrect, it is not considered a prompt.

56. b
By using multiple baseline design across behaviors and introducing each behavior at different times, you can determine if DRO is effective.

57. c
As a BCBA, you must consider other least restrictive interventions and explain the negative side effects of applying corporal punishment.

58. a
Procedural fidelity (AKA: treatment integrity) means that the independent variable is implemented as planned and there are no unplanned variables involved.

59. c
Because the access to the reinforcer (TV) for all the children depends on Kyle's behavior, it is dependent group contingency.

60. b
In a behavior cusp, learning a new behavior opens a new environment for an individual and it lets him/her encounter new contingencies.

61. a
Even though Nick agrees to follow the schedule, it is his decision whether he smokes or not. The behavior plan does not have the authority to prohibit someone from engaging in a behavior. It is rather a support to live better.

62. c
Not raining decreases the effectiveness of using the umbrella and rain boots; therefore, it is an AO.

63. d
Teaching loosely randomly changes the environmental setting while you teach a new skill.

64. c
Component analysis tests two or more independent variables to identify one that has a functional relation.

65. b
In discontinuous measurement, there are three types of time sampling: whole interval recording, partial interval recording and momentary time sampling, and these over- or underestimate the rate of the target behavior because of the way it is measured and it is called an artifact.

66. c
When Andrew is engaging in other behaviors but not hitting himself, the aversive stimulus (the sound of the vacuum cleaner) is removed; therefore it is DNRO (Differential Negative Reinforcement of Other behaviors.)

67. a
PRT (Pivotal Response Training) was developed by Robert Koegel and Laura Schreibman and it targets self-initiation of the subject's preferred activities in a natural setting.

68. c
If the replacement behavior is not in his/her repertoire, there is no behavior to reinforce.

69. d
Sign language, reading, writing and making vocalization are all verbal behaviors.

70. a
By implementing NCR (noncontingent reinforcement), the effectiveness of a reinforcer decreases; therefore, it functions as an AO (abolishing operation.)

71. c
In a sensory extinction, the problem behavior still occurs, yet it would not produce automatic reinforcement; therefore the individual would eventually stop engaging in the behavior.

72. b
When it is fixed, people will learn when their behavior will be reinforced; therefore, both FI (fixed interval) and FR (fixed ratio) tend to have a post-reinforcement pause.

73. c
Magnitude is the change in intensity or force of the response. It includes physical strength and volume of the voice.

74. b
A bootleg reinforcement occurs when one receives a reinforcer without meeting the predetermined criterion.

75. c
A naive observer is a person who collects data without knowing the purpose of the study.

76. b
Extinction is applied after a behavior occurs; therefore it is not an antecedent intervention.

77. d
In shaping, a response is differentially reinforced contingent upon the successive approximation toward an ultimate goal.

78. b
When a new skill is introduced, continuous reinforcement (CRF) should be used. When the subject learns the skill, an intermittent schedule of reinforcement should be used.

79. a
Respondent extinction occurs when a conditioned stimulus (the door knob) is no longer paired with an unconditioned stimulus (the squeaking sound) and the conditioned stimulus stops eliciting the response (getting goose bumps).

80. c
The song was added as a consequence (positive) and the behavior (clapping his hands) decreased (punishment); therefore, it is positive punishment.

81. d
When creating a goal, we need to make sure that learning opportunities are present. Ryan could behave well without tantrums for more than five days if non-preferred food were not on the table.

82. a
Verbal praise is learned; it is not an unconditioned reinforcer.

83. a
Indirect measurement provides subjective information regarding the target behavior. It may not be valid but it can be used to gather useful information.

84. c
Because a loud noise is removed (negative) and the behavior (doing her homework) increases (reinforcement), it is negative reinforcement. Because she has her brother do it, it is socially-mediated. Therefore, this scenario represents social negative reinforcement.

85. b
Contingent exercise is a punishment procedure and requires a person to engage in a behavior that is topographically different from the problem behavior.

86. c
It is important to understand the limit of our practice and seek appropriate treatment by professionals in other fields.

87. a
Because what Natalie is measuring (correct spelling) is not relevant to what Alvin is asking her to measure (speed of writing), it is not valid.

88. b
Three levels of scientific understanding are description, prediction and control.

89. a
The dependent variable is the quantitative measurement of the target behavior.

90. d
In the "applied" dimension, the target behavior must be socially important and benefit the individual's life. In this scenario, teaching Michelle to follow his brother's requests would not benefit her life.

91. c
PLACHECK (planned activity check) is a recording method on a group behavior and the observer headcounts the number of people engaging in a target behavior at the end of each interval.

92. b
Formal similarity requires two behaviors to resemble each other and have the same sense (eg., visual or auditory).

93. c
A behavior trap occurs when a person receives a reinforcer after engaging in a simple response and it quickly remains in his/her repertoire for a long time.

94. a
Scatterplot is done in a direct observation.

95. c
Chaining with task analysis would teach her the key steps of using the washing machine. Premack Principle and a token economy are used to promote motivation. Motivation is not an issue in this scenario.

96. d
Extinction for a behavior maintained by automatic reinforcement (sensory extinction) is done by masking the delivery of reinforcement. Melissa can still engage in the same behavior, but she would not get the same reinforcement.

97. d
Spontaneous recovery occurs when a behavior (pressing the eject button) reappears after the behavior has decreased due to extinction.

98. a
To get the percentage of unscored-interval IOA, you must look at unscored data (non-occurrence) and divide 2 agreed intervals (interval 5 and 6) by 5 total intervals of agreed and disagreed intervals (interval 2, 5, 6, 8 and 10) and multiply by 100. (2/5x100=40%)

| Interval # → | 1 | 2 | 3 | 4 | 5 | 6 | 7 | 8 | 9 | 10 |
|---|---|---|---|---|---|---|---|---|---|---|
| Observer 1 | X | X | X | X | O | O | X | O | X | O |
| Observer 2 | X | O | X | X | O | O | X | X | X | X |

99. b
A false negative occurs when one concludes that the independent variable has no effects on the dependent variable even though it actually has had an effect.

100. a
Private events are covert (not observable) behaviors.

101. d
Reactivity is demonstrated when the subject is aware that he/she is observed and he/she reacts differently.

102. b
Stacy's statement should not be subjective. Empiricism requires a person to conduct an objective observation of a target behavior.

103. a
The time delay is an antecedent response prompt; therefore, it exists between the presentation of the $S^D$ and the prompt.

104. c
When the rate requirement is too high, a subject may stop engaging in the target behavior, which is called ratio strain.

105. c
A cumulative graph data never descends.

106. c
Even though Kim is familiar with an FBA, she is not a certified behavior analyst and therefore it should be conducted under the BCBA supervision.

107. d
Default technologies refer to the use of coercive and punitive interventions without elaborating the use of other effective interventions.

108. b
External validity focuses on generality to other environmental settings.

109. b
Tokens are considered valuable because they have backup reinforcers.

110. c
A contingent observation is a type of time-out procedure. It withholds the opportunity to engage in an activity by removing the subject from the activity setting and have him/her observe others engaging in the activity.

111. a
Giggling is an involuntary behavior; therefore, it is a respondent behavior.

112. b
Stimulus discrimination requires one behavior (yelling to get a cigarette) and two stimuli (Mary and Robert). The behavior is reinforced under the presence of one stimulus (Mary) and not reinforced under the presence of another stimulus (Robert).

113. a
Transitivity is a part of stimulus equivalence. It shows stimulus relations with A=B, B=C; therefore, A=C. (A: Roommate, B: Brazilian, C: Portuguese)

114. b
X axis is known as abscissa or horizontal axis.

115. a
Response latency is the time between an $S^D$ and a response.

116. d
By withholding Mark's preferred activity, his mother is depriving Mark of it so it would become a strong reinforcer to engage in the target behavior. The deprivation serves as an EO.

117. b
Multiple baseline design is conducted across subjects, behaviors and settings.

118. d
Double-blind control occurs when neither the subject nor the researcher can identify the presence or absence of the independent variable. It is used to eliminate a biases perception.

119. a
Due to the past aversive event, the toddler is crying upon seeing the doctor. Because it is a learned behavior, the doctor is a conditioned stimulus.

120. c
Both the written word, "dinosaur" and verbalizing the word "dinosaur" start and end the same; therefore, it has a point-to-point correspondence.

121. c
Because the train comes every 20 minutes, it is fixed interval (FI). Reinforcement (door being open) is available for 15 seconds after the elapse of the interval; therefore, it is limited hold 15-second.

122. b
When there is a high degree of variability, it is difficult to determine a functional relation between the independent variable and the dependent variable.

123. a
Javier's behavior (running away) is verbally taught by his father and it is not learned through a consequence; therefore, it is a rule-governed behavior.

124. b
Overshadowing occurs when a subject cannot engage in a behavior due to an interruption by another stimulus.

125. d
In imitation, the model and the imitative behavior have to look identical (formal similarity) and the imitative behavior has to occur immediately after the model is presented.

126. b
The target behavior is cleaning the tennis balls and it increases; therefore it is a reinforcement procedure. It increases in order to remove the aversive stimulus (100 swings); therefore, it is negative reinforcement.

127. d
In an interdependent group contingency, everyone gets a reward when all of them meet the criterion.

128. c
Greg sneezing after breathing pepper in his nose would be an unconditioned response as it is an involuntary behavior.

129. b
A bedtime story manipulates his behavior; therefore, it is the independent variable.

130. a
In an explanatory fiction, the explanation of the cause and effect does not contribute to the understanding of the phenomenon.

131. d
The bear's cage is a conditioned stimulus (CS) as the incident changes the response to the presentation of the cage. Before the incident, the bear's cage would be a neutral stimulus.

132. a
Temporal locus recognizes that all behaviors have a starting point.

133. d
In a four-term contingency, there are EO, $S^D$, Response and Reinforcement.

134. b
You should start with continuous reinforcement, move to a fixed ratio and change to a variable ratio (intermittent schedule of reinforcement.)

135. c
In a contingency independent antecedent event, rather than a consequence, the MO (being hungry) affects the behavior.

136. b
When Elisa is playing the videogame, her hands are occupied and she cannot bite her nails simultaneously. Elisa playing the videogame is reinforced while biting is absent; therefore, it is DRI (differential reinforcement of incompatible behavior.)

137. c
Because the toy phone does not work, reinforcement would not occur when John answers it (extinction); therefore, it is an S-Delta.

138. c
Trials to criterion can be measured by checking the record of trials for the target behavior needed to achieve the preset criterion.

139. c
A Standard Celeration Chart has a slope of 34 degrees in its chart.

140. c
In momentary time sampling, the behavior is checked when the interval ends.

141. a
A changing criterion design is not an ideal method when the behavior is not in his/her repertoire. To see the progress of newly acquired reading words, a cumulative graph should be used instead.

142. d
Response latency is the time between an $S^D$ and a response. Interresponse time is the time between two responses in the same response class.

143. c
When using an exact count-per-interval IOA, the number of intervals that have the same frequency are divided by the total number of intervals. There are ten total trials and six intervals having the same frequency; therefore, it is 60%. (6/10x100=60%)

| Interval # → | 1 | 2 | 3 | 4 | 5 | 6 | 7 | 8 | 9 | 10 |
|---|---|---|---|---|---|---|---|---|---|---|
| Moe | //// | /// | /// | // | //// | / | /// | // | 0 | /// |
| Gloria | // | /// | /// | // | //// | 0 | /// | /// | 0 | // |

144. b
If the intensity starts low, the subject may be accustomed to the punishing stimulus and it may not impact the behavior.

145. b
BCIS (behavior chain interruption strategy) is used to confirm that the subject can emit an appropriate alternative behavior when one of the steps in the chaining is interrupted.

146. b
In permanent product, the outcome of the target behavior is observed and the change in the environment is measured.

147. c
Feedback should be provided immediately after the training session.

148. d
In parametric analysis, the effect of the varied degree of the same independent variable is compared.

149. b
In arbitrary stimulus class, stimuli do not look alike but share the same response.

150. b
Empiricism focuses on objective observation of a precisely defined behavior. This scenario only includes interviews, not objective observation. Furthermore, bad behaviors are not operationally defined.

151. a
Both fixed ratio and fixed interval show a post-reinforcement pause but fixed ratio has a high, steady rate.

152. d
She would stop her behaviors when situations cease; therefore these are all examples to escape situations.

153. d
George's behavior (hitting his classmate) has increased after receiving the consequence of his mother picking him up; therefore, it is reinforced by the consequence.

154. d
Hero procedure is also called a dependent group contingency.

155. d
When a reversal design is used, the baseline condition should be short for ethical reasons.

156. c
A line graph is the most commonly used format in ABA. It shows variability of target behaviors across time.

157. b
In a behavior chain with a limited hold, a person has to complete the required task within a limited time in order to produce reinforcement.

158. b
The required elements in baseline logic are prediction, verification and replication.

159. d
Because the child's self-injurious behavior is physically prevented and he cannot complete the behavior, it is response blocking.

160. c
A threat is not a punishment unless the consequences are actually delivered or the threat itself is the consequence and it decreases the behavior.

Made in the USA
San Bernardino, CA
05 September 2014